1

Inventions

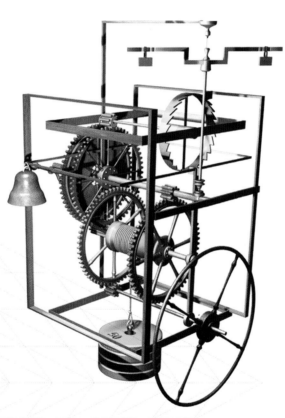

A TEMPLAR BOOK
First published in the UK in 2008 by Templar Publishing,
An imprint of The Templar Company plc,
The Granary, North Street,
Dorking,
Surrey,
RH4 1DN
www.templarco.co.uk

Conceived and produced by Weldon Owen Pty Ltd
59-61 Victoria Street, McMahons Point
Sydney, NSW 2060, Australia

Copyright © 2008 Weldon Owen Pty Ltd
First published 2008

WELDON OWEN GROUP
Chairman John Owen

WELDON OWEN PTY LTD
Chief Executive Officer Sheena Coupe
Creative Director Sue Burk
Concept Development John Bull, The Book Design Company
Publishing Coordinator Mike Crowton
Senior Vice President, International Sales Stuart Laurence
Vice President, Sales and New Business Development Amy Kaneko
Vice President, Sales: Asia and Latin America Dawn Low
Administrator, International Sales Kristine Ravn

Project Editor Lachlan McLaine
Designer John Bull, The Book Design Company
Art Manager Trucie Henderson
Illustrators Leonello Calvetti, Godd.com (Markus Junker, Rolf Schröter, Patrick Tilp),
Malcolm Godwin/Moonrunner Design

ISBN: 978-1-84011-727-1

Colour reproduction by Chroma Graphics (Overseas) Pte Ltd
Printed by SNP Leefung Printers Ltd
Printed in China 5 4 3 2 1

A WELDON OWEN PRODUCTION

insiders

Inventions

Glenn Murphy

45.58 kms

templar publishing

Contents

introducing

in *focus*

in_troducing_

What Is an Invention?

The word "invent" comes from the Latin for "to find or come upon". But invention is very different to discovery. Discovery is uncovering something that already exists, but invention means creating something completely new that never existed before. Can you imagine living without lightbulbs and washing machines? We would be reading by candlelight and scrubbing our clothes clean every night. But these things did not exist simply to be discovered. Instead, somebody, somewhere, at some time had to invent them. Inventions can change the world almost overnight, like the World Wide Web. Other inventions may not catch on for centuries, such as the submarine, which was invented in 1620 but didn't come into wide use until 1900. Look around and consider how many things we use every day that had to be invented.

TEAMWORK

In the past, great inventions were very often the work of a single person with one idea and everything he or she needed to make it real. As technology has become more complex and the problems we face more difficult, nowadays major inventions are coming more often from large teams of scientists or engineers working together.

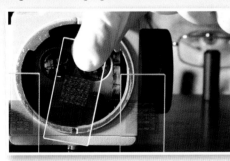

Array of hope
Researchers can now use DNA microarrays or "chips" to experiment with thousands of genes at once, where before they could deal with only one or two at a time.

Soaring genius

More than just a great artist, Leonardo da Vinci (1452–1519) was also a brilliant engineer and inventor. Many of the machines he first designed would not become a reality until hundreds of years after his death. Among these were designs for a parachute (invented in 1783), hang-glider (1891), tank (1900), helicopter (1907) and Aqua-Lung (1943).

Flying machine *People dreamed of flying for thousands of years before it became a reality. Leonardo da Vinci designed many flying machines modelled on the wings and flapping actions of birds and bats. Unfortunately, few would have flown.*

Timeline of Invention

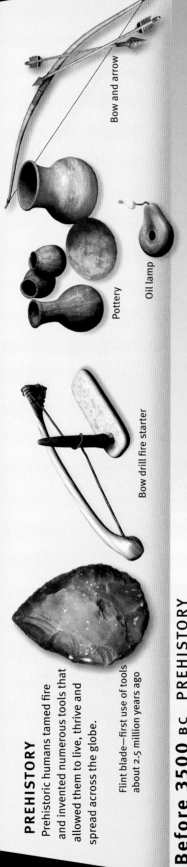

PREHISTORY

Prehistoric humans tamed fire and invented numerous tools that allowed them to live, thrive and spread across the globe.

Flint blade—first use of tools about 2.5 million years ago

Bow drill fire starter

Pottery

Oil lamp

Bow and arrow

Before 3500 BC PREHISTORY

BRONZE AND IRON AGES

People developed new tools such as, writing, metalworking, wheels and sails that enabled them to form the first civilisations, and explore the lands and seas.

The first wheels—Sumer

Writing

Metal swords and armour

Sailing boat

3500–500 BC BRONZE AND IRON AGES

THE ANCIENT WORLD

The ancient civilisations of Greece, Rome, China and the Middle East used wind, water and machine technology to build their powerful empires.

Water wheel

Chinese ocean-going junk

Hero's steam engine

Abacus

500 BC–AD 500 THE ANCIENT WORLD

THE MIDDLE AGES

China and the Islamic world were the great centres of invention in the Middle Ages. In Europe, the first printing machines fuelled an explosion of information and learning.

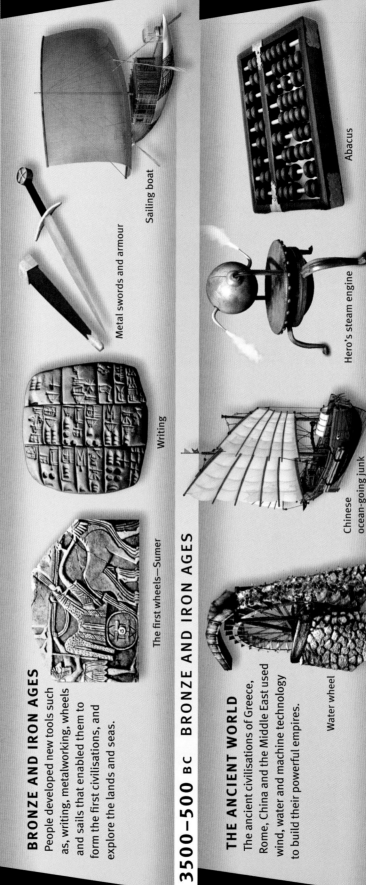

Gunpowder

Printing press

Drebbel's underwater rowing boat

Pascal's mechanical calculator

Microscope

Newton's reflecting telescope

THE SCIENTIFIC REVOLUTION

A revival of interest in scientific study spurred the invention of many precision instruments and set the stage for centuries of accelerated technological progress.

1450–1770 THE SCIENTIFIC REVOLUTION

Babbage's difference engine

Camera

Steam locomotive

Hot-air balloon

THE AGE OF MACHINES

From the late 1700s onwards, European inventors became engineers—finding new ways to power vehicles and manufacture goods. Engines and machines changed the world forever.

1770–1870 THE AGE OF MACHINES

Radio

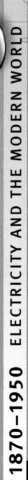

Wright Flyer

Telephone

Automobile

ELECTRICITY AND THE MODERN WORLD

The early 1900s saw three inventions that would change the world. Cars and aeroplanes made the world seem a smaller place, while electrical generators powered it in a whole new way.

1870–1950 ELECTRICITY AND THE MODERN WORLD

Maglev train

Bionics

Rocket

Electronic computer

THE DIGITAL AGE

Digital technology breaks down information into a series of numbers and processes it at lightning speed. From the mid-1900s, digital computers and electronics have revolutionised our lives.

1950–Today THE DIGITAL AGE

The First
Inventors

While animals use their teeth, claws and instincts to survive, it is the ability to make tools that sets us humans apart. The earliest tools—including spears, axes, arrows and fishhooks—enabled our Stone Age ancestors to survive as hunter-gatherers and spread across Earth. Later inventions—such as ploughs, sickles, pottery and buildings—allowed people to farm crops and herd animals. Now humans could create their own food sources instead of chasing them. In many parts of the world this change led to permanent settlements in villages, towns and eventually cities that would become the centres of new civilisations.

Stone Age tool kit

By 10,000 BC our ancestors had invented a wide range of methods and tools to help them survive and to make life less difficult. The most important of these were tools for making fire. Fire could be used for heating and for defence against predators and to harden spear points, cook food, bake bricks and clear grassland for farming.

Bow drill *Our early ancestors did not know how to make fire. Instead, they had to capture it from fires sparked by lightning strikes and then preserve it as burning wood or hot embers. Later, people invented tools to create fire whenever it was needed. The bow drill spun a sharpened wooden spindle to create heat by friction that set fire to the tinder around it.*

Cooking *The first cooks roasted meat on sticks over open fires. Heat makes meat safer to eat by killing bacteria. Evidence suggests that our ancestors were cooking with fire as long as 1.4 million years ago.*

Art *Prehistoric artists were painting cave walls and carving figures as long as 32,000 years ago. Surviving Stone Age art tells us much about the lives and beliefs of the people who created it. Many early carvings depict gods and goddesses, while cave paintings often show hunting and battle scenes.*

1. Woven cloth
2. Jewellery
3. Clay pottery
4. Flint axe head
5. Flint fishing spear
6. Bone fishing spear
7. Oil lamp
8. Musical instruments
9. Bone needles
10. Spinning twine
11. Axe
12. Bow
13. Woven basket
14. Arrow

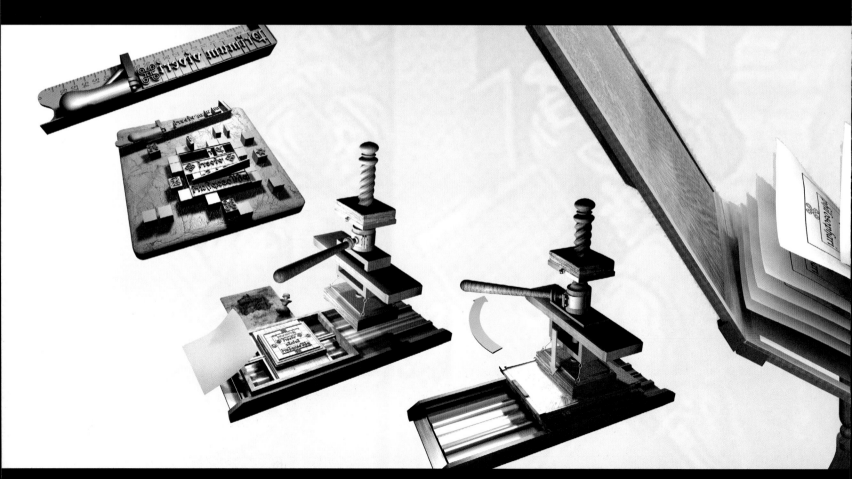

in *focus*

It's Revolutionary
Wheel

The wheel seems like such a simple invention that it is easy to forget how ingenious and important it really is. Forming a disc or cylinder and having it turn on an axis creates a simple device that serves many purposes. Fixed to a load or a vehicle, wheels decrease the drag caused by friction, allowing people and goods to be moved quickly over long distances. In this way, the wheel has helped shape human history for thousands of years. As part of a machine, the rotary motions of wheels and gears have powered everything from clocks to steam engines, and from aeroplane propellers to computer disk drives. In fact, you could say that wheels make the world go round.

The evolving wheel

The basic function of the transport wheel has not changed in over 5,000 years—it allows a load or vehicle to move freely across a surface. But on its journey from simple log roller to high-performance car or motorcycle wheel, the wheel has become stronger, lighter and more complex—giving a faster, smoother ride.

Spoked wheels *By Roman times wheels had become wider, lighter and more efficient. Weight was removed from the middle by creating spokes that connected the central hub to the outer rim.*

Sumerian wheels *These wheels were made from three planks of wood, pegged together and carved into a circular shape.*

Log rollers *In ancient times people rolled loads into place on logs. With use, the parts of the log in contact with the loads would have worn away. The result would have been something like an axle with wheels on the end, which was much lighter and easier to use.*

c. 4000 BC Getting things moving
Placed directly under stone slabs or sledges, log rollers were used to shift the building blocks of Stonehenge around 3000 BC but were probably used thousands of years earlier.

c. 3500 BC The first vehicles
The first true wheels we know of were found on carts in ancient Sumer (today part of Iraq). By 2500 BC the Sumerians were driving fast, four-wheeled chariots drawn by horses.

c. 300 BC Empire on wheels
The Romans built roads throughout their vast empire to move their armies and trade goods. Some modern highways follow the lines of Roman roads.

Train wheels *Early train wheels had solid iron hubs, spokes and rims in order to support heavy steam engines and carriages. The driving wheels were linked together to transfer power from the engine.*

LIVING WITHOUT WHEELS

Centuries ago, entire civilisations thrived in North and South America without using wheels for transport. The Incas and Aztecs made wheeled toys for children—so why not make carts and wagons? Perhaps no one thought to use wheels in that way. Perhaps wheels were not practical in the mountains and jungles where these civilisations were located. We may never know.

Pneumatic wheels *Pneumatic tyres were first fitted to bicycles. They meant a better grip and a more comfortable ride. This invention spread quickly to cars and other vehicles.*

Motorcycle wheel *The strong, lightweight design of modern superbike wheels is the result of more than 5,000 years of wheel development. High-grip, flexible, compound rubber tyres grip the road tightly, and disc brakes around the hub help control speed.*

1800s Steam revolution
The steam train was a huge revolution in wheeled transport. Suddenly people and goods could travel long distances faster than a galloping horse.

c. 1890 Transport for all
Early bicycles were called "boneshakers", because they gave a very bumpy ride. But once pneumatic tyres were invented in 1888, bicycles became a popular means of personal transport.

c. 1920 World in motion
Today people move faster and more often than ever before. The modern world is completely dependent on cars, trucks and the roads they run on.

With the Wind

Sail

The first rafts and canoes were crafted from logs more than 12,000 years ago. Some time later, the ancient Arab and Polynesian peoples found that hoisting a sail to catch the wind could help overcome currents too strong for paddling or rowing. From there, boats evolved into huge warships and trading ships with massive sails that did the work of a hundred heavy oarsmen. From China to Arabia and Polynesia, from Europe to Africa and the Americas, sailing ships spread goods, ideas and people to every continent.

NEW WAYS TO NAVIGATE

The first ships sailed along rivers and coastlines where land was in sight. Taking to the open seas meant losing sight of land, and the risk of never finding it again. These inventions use the sun, stars, magnets and satellites to help sailors find their way.

Persian astrolabe

Chinese compass

Sextant

Global Positioning System receiver

Battling behemoth

The HMS *Victory* was a ship-of-the-line in the Royal British Navy. It famously fought in the Battle of Trafalgar—the last great battle between sailing ships. Its masts held up to 37 sails, with a total area of 5,500 square metres (59,000 sq ft).

Main mast

Foremast

Fore topgallant studding sail

Main topgallant studding sail

Main topgallant sail

Main topsail

Mizzen topgallant sail

Mizzen mast

Main top studding sail

Foresail

Lower studding sail

Main trysail

Mizzen topsail

Spanker sail

Ropes and rigging *HMS* Victory *had over 42 kilometres (26 mi) of hemp and flax rope to hoist and control the sails. New crew had to literally "learn the ropes", which included tacks, sheets, clewlines, bowlines, buntlines, slablines, shrouds, ratlines, vangs, brails and halyards.*

c. 1850 Racing yacht
Diesel engines have replaced sail for passenger, cargo and fighting ships. But sailing for sport is more popular than ever. Modern racing yachts are built from light fibreglass.

c. 1650 Ship of the line
For centuries naval battles were usually won by the biggest ships, with the most guns. Ships of the line were the most powerful gunships around and the ancestors of modern battleships.

c. AD 200 Chinese junk
Chinese junks had sails kept flat with bamboo poles. Large, ocean-going junks sailed the world, bringing gunpowder and paper to Arabia and Europe.

c. 3500 BC Egyptian trading boat
Ancient Egyptian trading boats sailed up and down the Nile River and from there to the open seas of the Mediterranean.

TIMELINE
THE SAIL

Keeping Time
Clock

Perhaps no invention rules our lives as much as the humble clock. Clocks measure out the hours, minutes and seconds of our days from waking to sleep. But it was not always so. In ancient times, hunters and the first farmers tracked time by the Sun and stars, and the changing seasons provided their calendars. It was not until the first civilisations appeared about 5,000 years ago that people began splitting the day into periods. The ancient Babylonians and Egyptians created shadow clocks and sundials, measuring the hours by the movement of the Sun. Later, the Greeks and Chinese built water clocks that could work day and night. But it was in medieval Europe that the first fully mechanical clocks—powered by weights or springs—were developed.

Keeping good time

The escapement is the part of a mechanical clock or watch that keeps it regular. It works by releasing energy from a falling weight or uncoiling spring at a regular rate. Without this, the irregular release of power to the hands would make them speed up or slow down over time. The first clocks used a mechanism called the verge escapement, which is illustrated here.

Foliot *The verge is topped by a crossbar called the foliot. Weights on each end of the foliot can be adjusted to regulate its swinging motion.*

Verge *The verge is a vertical rod suspended by a thread from above for free movement.*

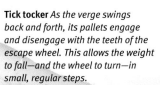

Weight *The power to drive the clock comes from a weight that falls under gravity.*

Tick tocker *As the verge swings back and forth, its pallets engage and disengage with the teeth of the escape wheel. This allows the weight to fall—and the wheel to turn—in small, regular steps.*

Escape wheel *The movement of the toothed escape wheel is checked by two flat projections (called pallets) attached to the verge.*

TIMELINE

THE CLOCK

c. 3000 BC Shadow clocks
The earliest clocks were shadow clocks and sundials, which measured time by the movement of a shadow around a fixed object. The ancient Egyptians used these to split their days into parts.

c. 270 BC Water clock
Early water clocks were simply bowls of water. Time was measured on a scale as the water drained through a small hole. Greek engineer Ctesibius added a second bowl with a float that turned an hour hand as it lifted.

c. AD 1000 Hourglass
Hourglasses or sand clocks were mostly used by sailors and navigators at sea. Neither sun nor water clocks worked properly on a moving ship, but a set of hourglasses could keep time well enough.

Clocks for the faithful

The first mechanical clocks were built so that churches and monasteries could hold services and prayers at regular times. Many clocks struck the hours on a bell. These early clocks had only an hour hand. Mechanical clocks were not accurate enough to measure minutes until the late 1500s, after the pendulum was invented.

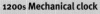

1200s Mechanical clock

Italian Giovanni di Dondi's amazingly accurate astronomical clock displayed the time and the date, and kept track of the movements of the Sun, Moon and planets. It was built within 100 years of the first simple mechanical clocks.

1972 Quartz watch

Canadian engineer Warren Marrison first used vibrating quartz crystals to regulate a clock in 1927. Quartz wristwatches followed by 1972. Today digital and quartz watches are powered by batteries, the Sun and even body movement.

Putting Wind to Work

Windmill

In medieval Europe, changes in farming techniques meant that a lot more grain could be grown with fewer peasants labouring in the fields. But with so many peasants moving from farms to towns, there were not enough people to manually grind the grain into flour to make bread. Instead, the millers turned to wind power. More than just mechanical grain grinders, windmills have also been used to pump water from wells and drain thousands of acres of land. Through this process, large parts of the Netherlands were drained that had previously been under water. Today, windmills can even "harvest" energy, using wind-powered generators to produce electricity.

Tower of power

Tower mills like this one were developed by the Dutch and English in the 1500s. This two-part tower design improved on the earlier post mills, as the miller did not have to turn the entire mill to face the wind—he simply turned the cap bearing the sails.

Sails *The sails are made of cloth stretched across a wooden grid called the sailback. Each sailback attaches to a stock, and working together, all four of these turn the windshaft behind them.*

Cap *This movable wooden dome sits on top of the stationary body. The cap carries the sails and houses the gear wheels attached to it.*

Brake wheel *This large wheel transfers power from the sails to the internal machinery. It also houses the brake used to slow or stop the mill.*

Wallower *This gear transfers the vertical motion of the brake wheel into the horizontal motion of the spur wheel.*

Spur wheel *This large central wheel transfers power to two smaller horizontal wheels on either side, each driving a heavy grindstone beneath.*

Grindstone *Not one, but two huge, circular stones mill the grain. The runner stone sits on top and turns with the shaft, while the bedstone beneath stays still.*

Gallery *A circular stage added to taller mills to allow the miller to adjust the sails or turn the cap into the wind.*

Spreading the Word

Printing Press

Before the mid-1400s, few people had access to written information, and most learned what little they knew from church sermons, town criers and gossip. Books were scarce, each copied by hand by monks or scribes. A few books, mostly on religious themes, were printed from wooden blocks that had to be painstakingly cut by hand—a slow, expensive method. All that changed after 1447, when Johannes Gutenberg of Mainz, Germany, invented the first movable-type printing press. Suddenly, printing became quick, simple and cheap. Books on all kinds of subjects—politics, philosophy, science—multiplied fast, fuelling the Renaissance, or "rebirth", of knowledge in Europe and elsewhere.

Frisket *This hinged wooden frame, attached to the movable schlitten, was used to lower the paper over the form for printing. It was covered in canvas or sheep's skin on each side, and could be padded with felt to give lighter or firmer impressions.*

● **Making a line** *Reading from a manuscript, the compositor arranged individual type characters on his composing stick, to create a line of text.*

● **Making a page** *Completed lines were slid onto a flat board called a galley, then locked together to create a form.*

● **Inking** *The form was placed on a sliding tray, transferred to the press and inked using handheld tools called ink-balls.*

● **Pressing** *Damp paper was placed on the form, and a heavy plate screwed down on top to press the type. The end result was a printed page.*

THE WRITTEN WORD

Written language was invented more than 4,000 years before printing. Many cultures drew pictures to represent words, but the ancient Sumerians went further. They spelled out their words using wedge-shaped symbols cut into clay to represent sounds.

The first writing This Sumerian clay tablet records a count of sheep and goats.

Books by hand Before Gutenberg, most books were written with ink and quill—a feather plucked from a goose with its end cut to a sharp point. Gutenberg's type letters were designed to look like this handwritten script.

THE PRINTING PRESS

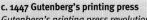

AD 1041 Chinese movable type
The Chinese were using block printing around AD 500, and printing whole books by 868. In 1041, 400 years before Gutenberg's press, alchemist Pi Sheng created the first movable type, made from baked clay glued to an iron plate.

c. 1447 Gutenberg's printing press
Gutenberg's printing press revolutionised printing almost overnight. Books, pamphlets and news sheets flooded through Europe, spreading knowledge and information to more people than ever before.

Clever copier

Gutenberg's invention made printing faster and easier by replacing whole-page printing blocks with reusable characters. The lettering could be moulded quickly, arranged in lines to make a page, then broken up and repositioned after printing to make new page templates.

Spindle *A big wooden screw called a spindle was turned with a lever to apply even pressure to the paper.*

Bridge *The spindle was guided through this heavy wooden crossbeam.*

Schlitten *A sliding frame called the schlitten held the bed of the press, allowing it to be moved back and forth for resetting and positioning under the press.*

Bed *The bed was the flat part of the press that held the form.*

Form *The form was coated in a sticky oil-based ink. One application could print up to 10 pages.*

1840s Rotary printing press

The next step in printing was to get machines to do most of the work. In 1810, steam-powered presses replaced the flat press with a metal cylinder. By 1847, improved rotary presses were producing 24,000 copies an hour.

1969 Office laser printer

First produced by the Xerox company in 1969, laser printers use electrostatic charges to direct ink onto rollers and from there to paper. By 1990, these rapid, accurate and compact printers could be found in offices and homes worldwide.

Telescope

No invention has changed our view of the universe as much as the telescope. The first telescopes were simply two glass lenses placed at either end of a tube—the earliest known was built by Dutch spectacle maker Hans Lipperhay in 1608. Soon afterwards, Italian scientist Galileo built an improved telescope and with it discovered moons around Jupiter and evidence for the idea that Earth orbits the Sun (rather than the other way around). Since then, telescopes have become much more powerful and the development of reflecting, radio and space telescopes has enabled us to delve ever deeper into the universe and to shed light on its origins.

James Webb Space Telescope (JWST) *The JWST is an orbiting space telescope due to be launched in 2013. With its larger mirror, infrared detectors and solar shielding, the JWST will look further into the universe than its predecessor, Hubble.*

MICROSCOPE

While the telescope revealed faint, distant stars and planets, the microscope showed the tiny features of insects, bacteria and cells invisible to the naked eye. Invented by Dutchman Zacharias Janssen and his father, Hans, in 1590, it was popularised by English scientist Robert Hooke. His 1665 book *Micrographia* showed fleas and other tiny objects magnified to fill a whole page, and inspired people across Europe.

Mirror vision

The Hale Telescope is based at Palomar Observatory in California. With its 5-metre (200-in) mirror, it was the largest reflecting telescope in the world when it was completed in 1948, and remained so for more than 30 years. In that time, astronomers used it to prove, among other things, that the universe is getting bigger.

Changing focus *Different viewing points around the telescope give different fields of view. One view is wide enough to see the whole Moon at once; another reveals just a tiny patch of sky in great detail.*

Coudé focus *A diagonal mirror can bounce light out of the telescope at right angles, into a spectrograph room for analysis.*

Prime focus *Here, the astronomer actually sits inside the telescope and makes direct observations from light reflected and focused by the primary mirror.*

Primary mirror *This huge mirror gathers as much light as possible and is the heart of the Hale Telescope.*

Elevator *Astronomers reach the observer's cage by riding this elevator.*

Cassegrain focus *Here, light is reflected back through a hole in the primary mirror to observers or instruments beneath the telescope.*

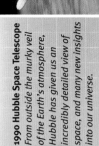

Powering the Modern Age
Engine

Engines are machines that convert energy from fuel into motion. Engines drive this conversion process so rapidly, they generate enough power to run another machine or vehicle. The first practical engines burned coal to boil water, generate steam and drive a moving piston. Steam engines powered the machines and vehicles of the 1800s and early 1900s, causing a revolution in industry, manufacturing and transport. Internal combustion engines work even more quickly to convert energy from fuel into motion. These engines ignite high-energy fuels under pressure inside their pistons to produce an explosive driving force for cars, boats and aeroplanes. Steam, however, is not as old-fashioned as you might think. Most of the electricity that gives us power today is produced by huge, steam-driven machines called turbines.

Engine of industry

By 1850, compact, high-pressure steam engines like this one had become the driving force for every mill and factory in the industrialised world.

Boiler *The boiler was heated with wood or coal. Water piped inside quickly turned to hot pressurised steam.*

PUTTING STEAM TO WORK

Steam engines were first used to pump water from mines and to power factory machines. Soon afterwards, they began to be used for transport and eventually replaced horse power on land and wind power at sea.

Steam locomotive

Paddle steamer

Steam tractor

TIMELINE

THE ENGINE

c. AD 60 Hero's aeolipile
Greek engineering genius Hero built the first steam engine almost 2,000 years ago, but no use was made of it. Water inside a sphere was brought to a boil. Steam then jetted from two nozzles, rotating the sphere at high speed.

1712 Steam engine
Englishman Thomas Newcomen's steam engine of 1712 was built to pump water from mines. James Watt and Richard Trevithick improved the design and revolutionised world transport and industry.

1859 Internal combustion engine
*Belgian inventor Étienne Lenoir converted a steam engine to run o
coal gas ignited inside the piston
German Nikolaus Otto later impro
this design, creating a powerful e
with four pistons firing in sequenc*

Piston power

The double-acting steam engine was invented by Scotsman James Watt in 1782. Pressurised steam (red) from the boiler was piped to the cylinder via a sliding valve that delivered it alternately to each side of the piston. Exhaust steam (blue) left the engine through the steam outlet.

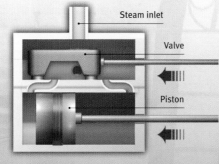
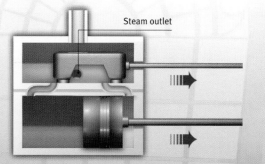

Steam inlet

Steam outlet

Valve

Piston

Governor *The engine's speed was regulated by the governor. If the engine started to work too fast, the weighted arms flew out, reducing the amount of steam delivered to the engine. If it slowed down, the arms dropped and the flow of steam increased.*

Flywheel *The heavy flywheel constantly turned, smoothing out the delivery of power to the machines.*

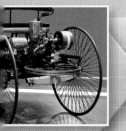

1892 Diesel engine

German engineer Rudolph Diesel designed a second type of combustion engine in which the air-fuel mixture inside the cylinders is compressed until it explodes. Diesel engines are particularly suited to powering heavy vehicles, tractors and boats.

1937 Jet engine

Jet engines work by forcing air into a tube where it is compressed, mixed with fuel, burnt and shot out the other end to produce thrust. British air force engineer Frank Whittle built the first working jet engine in 1937.

Capturing the Moment
Camera

The ancient Chinese and Greeks understood the principle of the camera more than 2,500 years ago. They discovered that light passing through a small hole in the wall of a darkened room casts an upside-down image of the outside world on the opposite wall. Such rooms were given the name *camera obscura*, meaning "darkened chamber". In the 1820s, Frenchmen Joseph Niepce and Louis-Jacques-Mandé Daguerre invented a way of capturing these images permanently on light-sensitive plates inside sealed boxes. By 1890, plates were replaced by rolls of celluloid film, and small, portable cameras could take 100 reprintable pictures. Today photography is easier and quicker than ever. We can capture, store, upload and e-mail images using digital cameras and cellphones.

Coded pictures

Digital cameras use lenses to focus images and shutters to control the time of the exposure, but unlike traditional cameras, they have no plates or film. Instead, they capture, store and display pictures by recording light and colour as digital information.

Digital code *Each CCD pixel translates the brightness of light into an electrical signal, converting the whole image into digital code. This code is processed by a tiny computer chip within the camera.*

CAPTURING MOTION

English photographer Eadweard Muybridge created the first moving picture or "movie" in 1877. Later he showed his idea to American scientist Thomas Edison. Edison used the idea to build his kinetoscope, the world's first movie projector.

The first "movie"
Muybridge used 12 cameras triggered by a trip wire to capture the motion of a galloping horse. He later arranged the photographs around a spinning disc to give the illusion of movement.

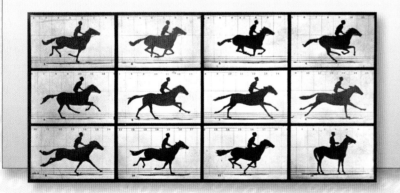

Display *The recorded image is instantly displayed on the liquid crystal display (LCD) screen on the back of the camera.*

TIMELINE

THE CAMERA

c. 500 BC Camera obscura
"Pinhole" cameras were invented in ancient times. Small portable versions became popular as artists' sketching tools in the 1600s.

1826 Chemical photography
Frenchmen Joseph Niepce and Louis-Jacques-Mandé Daguerre took the world's first photographs, using huge box cameras with chemical-coated metal plates.

1860 Colour photography
Scottish physicist James Clerk Maxwell created the first colour photograph by combining separate images taken using red, green and blue filters. Digital cameras use a similar method today

Lens *The lens collects light coming in through the aperture—a hole with shutters—and focuses it onto the CCD surface.*

Lenses and focusing

Cameras have changed a lot over time but one thing that has remained constant are lenses. These are essential for capturing a sharp image quickly.

Pinhole Pinhole cameras simply let light in through a tiny hole. Few light rays can pass through the hole at once, so it takes a long time to take a picture.

Let in the light Making the hole bigger allows more light rays to enter. But because the rays now overlay each other, the image is blurred.

Lens The solution is to add a lens. This focuses the extra light rays from a set distance and creates a sharp image to be recorded by the film or CCD.

Bayer filter *The CCD pixels measure brightness, but not colour. Colour information is gathered by filtering the light through an array of red, blue and green filters called the Bayer filter.*

Charge Couple Device (CCD)
The CCD is a silicon chip the size of a thumbnail, covered with millions of individual light-sensitive cells called pixels.

Chemical freeze-frames

Early plate cameras captured images by projecting them onto metal or glass plates covered with light-sensitive chemicals. As the chemicals reacted, they left light and dark patches on the plate, forming permanent black-and-white images. Making prints and adding colour came later.

1895 Motion pictures
French brothers Louis and Auguste Lumière used Thomas Edison's projector design to create the first moving images projected onto a screen for public viewing. This marked the beginning of cinema.

1988 Digital camera
The technology for digital cameras grew out of space probes and satellites developed in the 1960s. The first true digital camera, the Fuji DS-1P, did not arrive until 1988.

Transport on Track
Train

The idea of putting wheeled vehicles on tracks for fast, smooth running is an ancient one. The ancient Greeks and Chinese carved grooves in their roads for carts. Mining carts have run on wooden or metal rails since the 1500s. But the development of the train as we know it began when wheeled vehicles on tracks started to be powered by steam engines. Slow and cumbersome at first, trains advanced quickly during the 1800s, and their impact was enormous. Distant cities suddenly seemed close together as trains shuttled goods, workers and vacationers between them, quickly and reliably. Today, trains remain unbeatable for carrying commuters and cargo.

Flying underground
Maglev—short for MAGnetic LEVitation— trains have the potential to revolutionise transport. They are lifted and propelled at incredible speeds by powerful magnets, making them faster, more energy-efficient and less polluting than trains with engines. This illustration shows a Swissmetro train. Swissmetro is a proposed network of high-speed maglev trains to link the cities of Switzerland and France.

Dig deep
Trains on the surface have to run around, through or over mountains and valleys. The Swissmetro maglev trains will run through tunnels dug between 60 and 300 metres (200–1,000 ft) underground, avoiding them altogether.

Drivers' view
In the maglev cockpit, one or two drivers will control and monitor the propulsion system via a panel of instruments and digital displays.

TIMELINE

THE TRAIN

1804 Locomotive
English engineer Richard Trevithick built the first locomotive by combining a water-pumping steam engine and a mining cart.

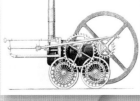

c. 1935 Streamliners
By 1900, trains had spread throughout Europe, Asia and America. In the 1930s aerodynamic steam locomotives called streamliners could cruise at speeds of 160 kilometres per hour (100 mph).

1940s Diesel electric
First used in locomotives in 1924, diesel engines were later coupled with electric motors to produce more efficient and reliable power for trains.

A WAY WITHOUT WHEELS

Magnetic levitation systems use two sets of magnets to lift and move the train. One set is attached to the underside of the train car, while a second set of switchable electromagnets lines the track or guideway. Once activated, these levitate, guide and propel the train using the forces of magnetic repulsion and attraction.

Guide magnets

Train magnets

Electromagnets in the guideway are switched rapidly back and forth between opposite magnetic poles.

The alternating guideway magnets attract and repel those on the train car. This both pushes and pulls the train forward.

1981 Train à Grande Vitesse (Very Fast Train)
The all-electric TGV is the world's fastest conventional train. TGVs began carrying passengers across France in 1981, and now run at 320 kilometres per hour (268 mph) throughout mainland Europe and to England.

2004 Magnetic levitation train
The world's first commercial high-speed maglev train service began running in Shanghai, China, in 2004. It runs between the city and its airport at a top speed of 430 kilometres per hour (268 mph).

Sound Ideas
Music Player

Permanent sound recording did for music what writing and printing did for spoken language. It allowed music, speech and sound to be recorded, kept and revisited at a later date. Beginning with Thomas Edison's first (crude) recordings, the music player transformed music by making it available to anyone at any time. Before this, all music was live music. But now a musical performance could be copied and distributed, helping to create the idea of home entertainment. But perhaps most importantly, recorded music has brought together different countries and cultures—united in their love of music from classical to hip-hop.

Trumpet *The wide, funnel-shaped horn or trumpet amplified the sound waves created by the diaphragm, making them loud enough to be heard.*

Get into the groove

The first record players were entirely mechanical. The turntable was powered by a hand-wound clockwork spring and the music was played without any electronic amplification. The sound waves of the music were recorded as shapes in the walls of a single spiral groove cut into each side of a record. The gramophone needle vibrated as it ran through the groove, and the vibration was passed to the diaphragm and trumpet to create and amplify sound waves.

Record *Until the 1950s records were made from a mixture of shellac—a resin obtained from beetles—various bonding agents and dyes.*

TIMELINE
MUSIC PLAYER

1877 Edison's phonograph
Thomas Edison originally created the phonograph to store telegraph messages. It recorded sound vibrations on a rotating wax cylinder. The cylinders quickly wore out during playback.

1888 Gramophone
Invented by Edison's rival, Emile Berliner, the gramophone substituted the recorded cylinder with a flat disc. The discs were hard-wearing and cheap to produce. This marked the birth of the recorded music industry.

1979 Walkman
Although magnetic recording tape was invented in 1928, this medium remained less popular than records until 1985. The Walkman stereo helped change this, by making recorded music personal and mo[...]

Diaphragm *The vibrations of the needle were transferred to a diaphragm. This created sound waves inside the trumpet.*

Stereo record player *The left and right channels of a stereo record are recorded on either wall of the record groove. As the stylus moves through the groove, it vibrates in two directions. Magnets in the cartridge above convert this vibration into an electrical signal.*

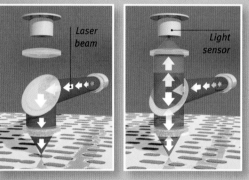

Laser beam

Light sensor

Compact disc *Music is encoded on the surface of a CD as a series of tiny pits. The pits are lined up in a single spiral and read by a laser beam, much like a record stylus tracks the groove of a record. If the laser beam hits a pit, it is reflected back to the laser (left). If it hits a space in between, it is reflected into a light sensor (right). These changes in reflectivity are converted into a digital signal that a computer turns into music.*

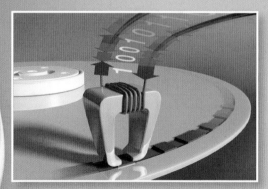

Digital audio player *In hard-disk digital audio players, music is stored as a magnetic digital code on a flat disk. This tiny disk spins thousands of times per minute and is tracked and decoded by a reader head using a magnetic field.*

Needle *The needle or stylus was made of very hard steel—or even sapphire or diamond—so that it would pick up every shape within the record groove.*

1982 Compact disc player
The compact disc (CD) was developed by engineers at the Philips and Sony companies. With no wear during playback and high-quality digital sound, CDs were outselling records by 1988 and cassettes by 1991.

1997 Digital audio player
Digital audio players store audio data on compact hard drives or flash drives. The Korean SaeHan company released the first digital audio player in 1997, four years before the hugely popular Apple iPod.

Making Electricity
Power Station

Using machines to produce power is not a new idea. In ancient times, waterwheels and windmills turned millstones, pumped bellows and beat hammers. But while these machines could transfer energy and power, they could not store it for later use, or deliver it for use in distant homes, offices and factories. For that, we needed electricity. With the invention of the electric generator in the 1800s, the world entered the new electric age. Most modern power stations rely on steam turbine generators powered by coal, gas or nuclear fuel. Others harness cleaner forms of power including hydroelectric, wind, solar, tidal and geothermal energy.

Electric waterwheel
At the heart of a hydroelectric power plant are turbine generators. The turbines convert the energy from rushing water into spinning motion. The generators—each consisting of a rotor and a stator—convert this rotary motion into electrical energy.

Stator *This stationary copper and iron magnet surrounds the rotor.*

Rotor *The rotor wheel is covered with electromagnetic iron cells. When it spins within the stator, an electric current is produced.*

Spiral casing *This pipe is shaped to generate maximum spin.*

Generator shaft *This transmits the spinning motion of the turbine to the rotor.*

Turbine *The water pushes the turbine blades as it rushes past.*

Heavyweight hydropower
Hydroelectric power stations harness the energy of falling dam water to generate electricity. They are some of the largest man-made structures on Earth and currently provide about six per cent of the world's electricity.

Crane *A crane is used to lift up the heavy generator rotors and turbines for maintenance.*

Transmission lines *The electricity produced by the generators is distributed to towns and cities along transmission lines.*

Inlet pipe *The water falls through the inlet pipes towards the turbines under the force of gravity.*

Reservoir inlet *The flow of water from the reservoir is controlled by opening and closing gates at the head of each inlet pipe.*

Turbine *The turbine is driven around by the force of rushing water. It is a modern version of the waterwheel.*

Generator *Connected to and driven by the turbines, generators rotate to produce electricity.*

Dam structure *Built from concrete reinforced with steel, its sheer weight holds back the massive body of water behind it. Its triangular shape helps prevent it from toppling.*

TIMELINE

THE POWER STATION

c. 200 BC Waterwheel
Waterwheels were used in ancient China, Greece and Rome to transfer the energy from running water to moving machinery.

1831 Electric generator
Englishman Michael Faraday discovered how to produce electricity using rotating magnets in 1831. One year later, French inventor Hyppolyte Pixii built the first practical electric generator.

1882 Hydroelectric power
The world's first hydroelectric plant was built on the Fox River in Wisconsin, United States. It powered two paper mills and a house.

1954 Nuclear power station
Nuclear energy was developed during World War II. The first nuclear power plants were opened in Russia and Great Britain in the mid-1950s.

Keeping Connected
Telephone

The word "telephone" means "distant voice", and that is just what this invention gave to humankind. Telephones carry our voices through wires, air and space, bridging the longest distances in an instant. Long-distance sound messaging—or telecommunications—began with the telegraph in 1837. But telegraph messages were usually brief and impersonal, as they had to be sent through trained operators—who tapped out code at one end and decoded the message at the other. Telephone networks developed quickly in the late 1800s; by 1887 there were more than 100,000 telephone subscribers worldwide. Today, with cellphones and satellite networks, we can chat with anyone, anywhere, at any time.

Miniature miracle

Thanks to rapidly developing microelectronic and microprocessor technologies, modern cellphones have become incredibly small, light and versatile. Some phones feature cameras, music players, Web browsers and more.

Routing calls

Making a call from one phone to another requires a complex sequence of signal transfers. Yet these happen so quickly and smoothly that we never even notice. A telephone number is like an address on an envelope: it tells the automated telephone exchanges where to direct the call. Cellphones send out continuous signals telling the networks where to find them should a call come through.

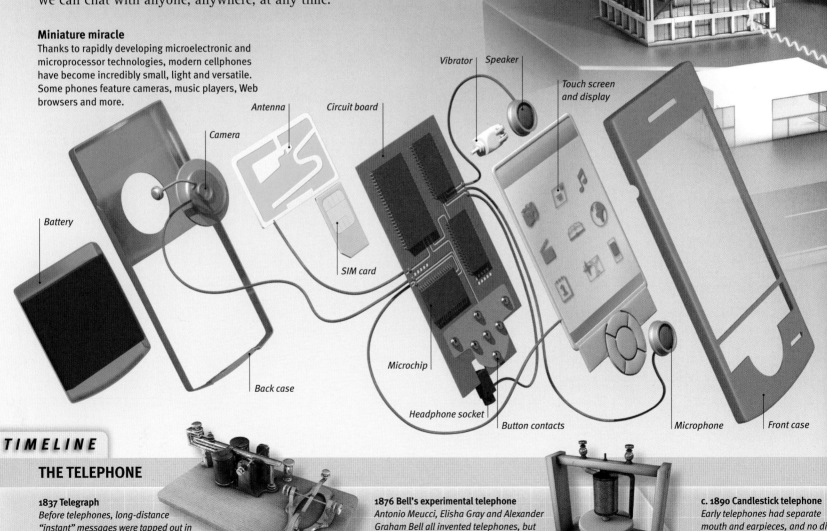

Labels: Battery · Camera · Antenna · Circuit board · SIM card · Back case · Microchip · Headphone socket · Button contacts · Vibrator · Speaker · Touch screen and display · Microphone · Front case

---

TIMELINE

THE TELEPHONE

1837 Telegraph
Before telephones, long-distance "instant" messages were tapped out in code and transmitted by telegraph. Practical electric telegraphs were invented by William Cooke and Charles Wheatstone in Britain, and improved by Samuel Morse in the United States.

1876 Bell's experimental telephone
Antonio Meucci, Elisha Gray and Alexander Graham Bell all invented telephones, but it was Bell's model that evolved into the modern telephone. His prototype used a diaphragm and a wire floating in acid to transform sounds into electricity and back again.

c. 1890 Candlestick telephone
Early telephones had separate mouth and earpieces, and no di buttons. The caller lifted the earp and gave a number to an operat working at a telephone exchang who would connect the call.

1 **Making the call** *An office worker dials a friend's cellphone number. The analogue signal is relayed to the local exchange.*

2 **Down the wire** *At the local exchange the call is converted to a digital signal for transmission by cable to the main exchange.*

3 **Through the air** *This call is routed by microwave link to the cellphone exchange. Other exchanges are linked by cable.*

4 **Cell-bound** *The signal travels by cable to the cellphone tower closest to the cellphone being called.*

5 **Connected** *The signal is relayed to the friend's cellphone and the call is answered. As the car moves, the signal is transferred smoothly from one cell tower to another.*

Cells explained
Cellphone networks divide the landscape around us into invisible cells. Each cell has a transmitter near the centre. Cells are smaller in built-up areas, giving the network greater capacity.

SUBMARINE TELEPHONE CABLE

Although many international calls are relayed by satellites, most go through undersea telecommunications cables that run for thousands of kilometres across deep ocean floors. Layers of metal and plastic waterproof the cable and protect it from damage by trawler boats and marine life. At the core is a thick bundle of optical fibres that carry digital telephone and Internet signals.

1919 Rotary dial telephone
The first dial telephones appeared in 1919, and from the 1930s onward mouth- and earpieces were built into one, hand-held receiver. Touch-Tone telephones followed in 1961, and soon almost every Western home had a telephone.

1978 Cellphone
The first cellphones were big, cumbersome and expensive, but rapid technological advances have since led to lightweight cellphones with built-in cameras, Internet browsers and digital audio players.

Deep Diver
Submarine

From their very beginnings, submarines were designed to be underwater warcraft, slipping beneath the waves to sink ships from below and turn the tide of naval battles. Some of the first submarines were used during the American Civil War in the 1860s. They relied on hand-powered cranks and pumps to propel and submerge them. Although they sank a few ships, they were more often useless and proved dangerous to their crews. Fifty years later, submarines with ballast tanks, diesel-electric engines and underwater torpedoes played a key role during World War I, sinking hundreds of ships and securing their place in the navies of the world. Along the way, submarines have taken on more constructive uses in deep-sea research, exploration and engineering.

IN DEEP

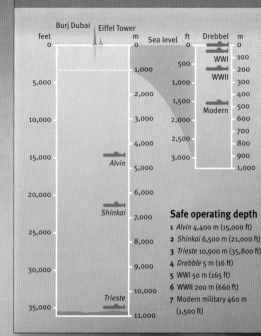

Submarines must be built to resist the crushing pressure of the water around them—a pressure that grows steadily greater the deeper they go. Research submersibles can go a lot deeper than military submarines. It is much easier to build a small super-strong chamber for two or three explorers than living and working space for dozens of crew. In 1960 the submersible *Trieste* with two crew members went to the bottom of the Mariana Trench—the deepest part of any ocean. No existing submersible can go as deep.

Safe operating depth
1 *Alvin* 4,400 m (15,000 ft)
2 *Shinkai* 6,500 m (21,000 ft)
3 *Trieste* 10,900 m (35,800 ft)
4 *Drebble* 5 m (16 ft)
5 WWI 50 m (165 ft)
6 WWII 200 m (660 ft)
7 Modern military 460 m (1,500 ft)

TIMELINE

THE SUBMARINE

1620 Drebbel
Dutch inventor Cornelius van Drebbel built the first working submarine for the British Royal Navy. His leather-coated, oar-propelled sub made several test runs in the River Thames.

1776 The *Turtle*
Built by engineer David Bushnell during the American Revolutionary War, the pedal-powered Turtle used naval mines to attack British warships in New York harbour. The attacks failed, but the military submarine was born.

1864 The *Hunley*
During the American Civil War, the Confederate army built several submarines to battle ships blockad their ports. The Hunley sank twice during practice runs before success sinking the Union ship Housatonic.

ALVIN

WOODS HOLE
OCEANOGRAPHIC
INSTITUTION

WOODS HOLE OCEANOGRAPHIC INSTITUTION
1930

OFFICE OF
NAVAL RESEARCH

MS 1512 M

Undersea explorer

Launched in 1964, *Alvin* is probably the world's best known submersible. It is still used by scientists to explore marine life and shipwrecks on the deep ocean floor. It can dive to depths of almost 4,500 metres (3 mi). *Alvin* has made over 4,000 successful dives, and has used its video cameras and robotic arm to explore the wreck of the *Titanic*.

1. Manipulator arm
2. Lights and cameras
3. Sail
4. Pressure hull
5. Viewport
6. Buoyancy spheres
7. Manoeuvring thruster
8. Batteries
9. Main propulsion unit

1916 U-Boat

With the addition of diesel-electric engines and torpedoes, submarines became much deadlier. During World War I, the German navy used them to sink hundreds of ships.

1948 Submersible

While submarines are designed for long, independent missions, submersibles make short, deep dives, assisted by a surface support ship. They are used for research and engineering projects on deep ocean beds and trenches.

ALVIN

Freedom Machine
Automobile

There were several innovations that led to the automobile as we know it today. In 1769, Frenchman Nicolas-Joseph Cugnot built the first successful steam-powered car, but its heavy engine could not manage more than a walking pace. The gasoline engine, invented in the 1880s, sped things up considerably. German inventors Otto Nikolaus and Gottlieb Daimler improved the engine and used it to build the first practical four-wheeled automobile in 1886. When American entrepreneur Henry Ford invented the car assembly line in 1913, cars became cheaper and simpler to build. His Model T "Tin Lizzies" sold in the millions. By 2000, the world's car manufacturers were building over 40 million vehicles each year, and the automobile had changed the face of the world forever.

Hybrid wonder

Where standard cars have just one engine driving the wheels, hybrid cars have two. In addition to a small, light gasoline engine, they also have an electric motor–generator powered by a stack of rechargeable batteries. The two engines combine and recycle energy, making them more environmentally friendly and cheaper to run.

Internal combustion engine

Energy management computer *The car's energy systems are constantly monitored and adjusted by a computer.*

Batteries and supercapacitors *These supply energy to the motor–generator when needed. They are recharged by the motor–generator at other times.*

Internal combustion engine (ICE) *The ICE helps to power the wheels when cruising at high speeds.*

Wheels *The wheels power the electric motor–generator when braking.*

Electric motor–generator *This powers the wheels when accelerating and at slow speeds.*

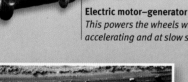

Driving our lives

Mass car ownership has changed the way we build our towns and cities. Millions of families now live in suburbs where leading a normal life would be impossible without owning at least one car.

TIMELINE

THE AUTOMOBILE

c. 2000 BC Horse and carriage
The horse and carriage was the most common form of wheeled transport for about 4,000 years.

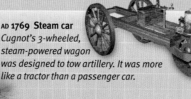

AD 1769 Steam car
Cugnot's 3-wheeled, steam-powered wagon was designed to tow artillery. It was more like a tractor than a passenger car.

1886 Daimler 4-wheeled car
Gottlieb Daimler built the first practical, gasoline-driven automobile. Its wooden wheels and steel tyres made it pretty rough to ride.

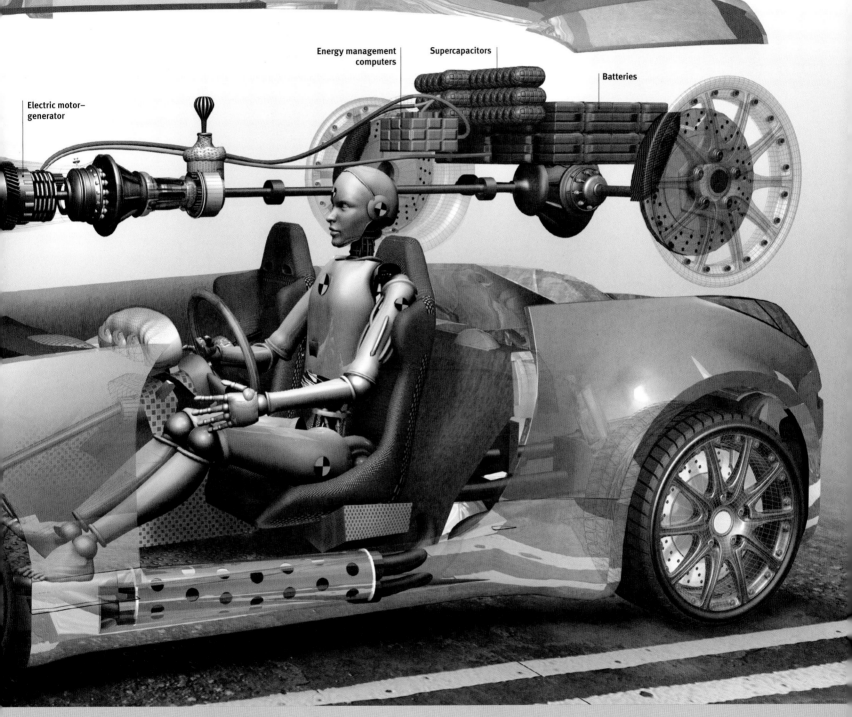

Energy management
computers

Supercapacitors

Batteries

Electric motor–
generator

1908 Ford Model T
*Built on rapid-assembly lines
after 1913, Henry Ford's
Model T helped to make the
automobile affordable for
millions of people worldwide.*

1938 VW Beetle
*The charming and reliable
Volkswagen Beetle (or Bug)
overtook the Model T's record
sales to become the most
popular car ever made.*

1997 Hybrid car
*Basic hybrid cars were built
and sold in small numbers
from about 1903 to 1920. The
modern hybrid car arrived
much more recently.*

Lighter Than Air
Airship

The first manned hot-air balloon flight lifted off in 1783, but it was another 100 years before anyone solved the major problem with lighter-than-air aircraft: how to fly against the wind. Early steam engines were either too heavy or too weak to propel a true airship, and even after lighter gasoline engines were invented, most airships were still too puny to carry passengers or cargo. Enter Ferdinand Graf von Zeppelin, who designed the first practical passenger airship in 1893, and started an age of luxury airship travel that lasted almost 40 years. Sadly, when the LZ-129 Hindenburg airship caught fire on 6 May, 1937, killing 35 passengers, people lost faith in airships, and the future of air travel was left to aeroplanes and helicopters.

Struts *The supporting girders were made of a lightweight aluminium alloy called duralumin. Steel cables were strung between them to keep the frame rigid.*

Framework *The airship's frame was constructed of 52 hooplike rings that were connected by 28 girders.*

Gasbags *Made of cow's intestines reinforced with cotton, these bags held in total about 105,000 cubic metres (3.7 million cu ft) of hydrogen gas.*

Engine pod *Engineers accessed the engine pods via external ladders.*

Propeller *The propellers were made from hard wood like ash and walnut. They had to be regularly replaced as they wore out.*

Engineer

Engine *The Graf Zeppelin had five engines, each delivering 550 horsepower. These pushed the zeppelin along at a speedy 130 kilometres an hour (80 mph).*

TIMELINE

THE AIRSHIP

AD 1783 Hot-air balloon
In November 1783, a science teacher and a soldier became the world's first aeronauts. They piloted a paper-and-cloth balloon made by French brothers Joseph and Étienne Montgolfier.

1852 Giffard's powered airship
Henri Giffard's hydrogen balloon flew at a top speed of about 10 kilometres per hour (6 mph) over Paris in 1852. But its steam-powered engine was too weak to propel it against the wind.

1892 Weather balloon
Weather balloons were used as early as 1892 to measure temperature and pressure in the upper atmosphere. They carry instrument packages that transmit data about conditions at high altitudes.

Skin *Added last during construction, the skin was made of cotton fabric coated with aluminium oxide paint. Tough but light, this gave the zeppelin a silvery colour.*

The floating hotel

The LZ-127 Graf Zeppelin was one of the most beautiful and admired aircraft in the history of the world. It was 236 metres (776 ft) long and 30.5 metres (100 ft) in diameter, and was built to carry 40 crew and 20 passengers. The Graf Zeppelin was the first passenger airship to fly from Europe to North and South America, and to circle the globe.

Gondola *This housed the crew and passengers, and had to be both luxurious and as light as possible. It had foam-and-fabric walls, and aluminium chairs so light you could pick them up with one finger!*

Navigation room

Radio room

Dining room

Passenger bathroom

Crew bathroom

Toilets

Control car

Kitchen

Passenger cabins

1898 Rigid airship
After the discovery of the lightweight metal aluminium, Graf von Zeppelin and others used it to build airships with rigid frameworks around gas cells that could be controlled independently, powered by lightweight gas engines.

GRAF ZEPPELIN

c. 1910 Blimp
Unlike zeppelins, blimps are nonrigid airships filled with high-pressure helium gas. They ascend and descend by letting out or taking in air, and are typically used for advertising and aerial filming.

Taking to the Skies
Aeroplane

Bird of prey *The F-22 Raptor is a swift and deadly jet fighter. It can maintain speeds and altitudes higher than those of any other fighter in the world.*

Although balloons and airships came first, today air travel is dominated by fast, reliable aeroplanes. Unlike balloons and airships, aeroplanes are heavier than air—aerodynamic forces allow them to take flight. At first, inventors tried to take to the air with pedal- or steam-powered flapping machines based on bird flight, but none succeeded. Pedal power was too feeble to keep a machine in the air, and steam engines were much too heavy. When the gasoline engine was invented in 1885, everything changed. It was light and powerful enough to be used in aeroplanes. On 17 December, 1903, the Wright brothers launched their clumsy, tailless biplane across the coastal sands near Kitty Hawk, North Carolina. A single gasoline engine propelled it a distance of just 37 metres (120 ft), but it made history as the first controlled, powered and sustained heavier-than-air human flight.

Double jumbo

As air travel has become more popular, jet airliners have had to grow in size to meet the demand. By 1970, the first 44-seater Comets had grown into Boeing 747 "jumbo" jets seating over 400. The Airbus A380 "superjumbo", however, can seat up to 853 people using upper and lower decks running the full length of the plane.

Control centre *The two pilots control the engines and flight surfaces with the aid of interactive computer displays, digital heads-up displays (HUDs), pop-up menus, sidesticks and trackballs. The primary computers are located below the cockpit.*

TIMELINE

THE AEROPLANE

c. 300 BC Kite
The first descriptions of kites, in China, date back more than 2,300 years. They established the basic principles of aerodynamics that later led to the inventions of gliders and aeroplanes.

1891 Lilienthal's Derwitzer Glider
German inventor Otto Lilienthal studied bird-wing shapes before building the world's first hang-glider. He made over 2,000 flights before dying in a gliding accident in 1896.

All aboard *The Airbus A380 has 50 per cent more floor space and 35 per cent more seats than a Boeing 747. The seats are arranged on two decks linked by stairs and elevators.*

Heavy roller *Twenty-two wheels under the wings, fuselage and nose help to spread the aeroplane's weight when it is on the ground. They fold under the wings or retract into the fuselage after take-off.*

Thrusters *At take-off, the A380's four jet engines produce more power than 3,500 family cars.*

1903 Wright Flyer
The world's first powered flight lasted just 12 seconds and reached the dizzying height of 3 metres (10 ft).

1943 Me 262
Although British engineer Frank Whittle invented the jet engine in 1938, it was the Germans who won the race to build the first practical jet-powered aircraft.

2007 Airbus A380
This colossal aeroplane is designed to carry as many passengers as possible, to ease traffic at airports and—in theory—make tickets cheaper. The first A380 entered service in October 2007.

Across the Airwaves
Radio and TV

Radio waves created by electric sparks were first detected in the 1860s. Radio as a means for communication was a reality some 30 years later. By 1929, radio and television programs were being broadcast to receivers across Europe and America. Now television and radio broadcasts reach billions—in many places they have replaced books and newspapers as the main sources of entertainment and information. In other uses, radio and television allow us to extend our ears, eyes and voices. Two-way radios let soldiers, pilots and air-traffic controllers talk to each other at a distance. Remote television cameras make it possible for surgeons to see inside bodies, and scientists explore Mars with robot rovers.

Breaking news

Before radio and television, news reports took hours, days or even months to reach their audience. Information had to be relayed by word of mouth, on paper or via telegraph signals. Now news—such as a sinking tourist boat—can be recorded live on location, relayed through land-based transmitters or satellites in space and brought to millions of people across the globe in a matter of seconds.

TIMELINE
RADIO AND TV

1895 Marconi's wireless telegraph
Italian inventor Guglielmo Marconi designed and built the first practical radio transmitter and receiver in 1895. Known as wireless telegraphy, the technology was first used to communicate with ships at sea.

1920s Consumer radio
Within a short time, wireless telegraphs developed into consumer radios. By the late 1920s, hundreds of radio stations were broadcasting information, music and entertainment to people in their homes.

1925 Baird's clever scrap box
Scottish inventor John Logie Baird built the world's first television, u[...] cardboard and scrap component[...] Incredibly, it worked, but poor pic[...] quality meant it was soon displac[...] by the first all-electronic televisio[...]

CAMERA TO SCREEN

During live outside broadcasts, the video signal is transferred instantly to a mobile broadcasting unit, then relayed to a television station via microwave signals or a satellite. From there, it can be sent to a transmission tower or satellite and beamed to televisions worldwide.

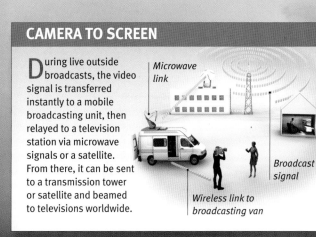

Microwave link

Broadcast signal

Wireless link to broadcasting van

Studio on wheels *The outside broadcasting van or "OB truck" is a mobile television studio, containing everything needed to record, edit and transmit live images and audio. Inside, the editor can switch between two or more cameras on the scene during the broadcast.*

1934 Electronic television
Working separately in the United States, Philo Farnsworth and Vladimir Zworykin built the first all-electronic television receivers and cameras. Within 30 years, 85 million television sets were sold in America alone.

1951 Colour televison
The first colour television program was broadcast in 1951, but very few people had sets on which to watch it. The latest sets have ultra high-definition flat screens and can receive digital signals via satellite and cable.

Quick Thinking
Computer

In many ways, the digital computer is the ultimate human invention. Much as telescopes extend our eyes and radios our ears, computers allow us to process information faster and more accurately than we could ever hope to do ourselves. The first computers were simple counting and calculating devices like the abacus. The first mechanical calculators appeared in the 1600s. Computers were improved—or rather, reinvented—during the last century, alongside developments in electronics. Within 10 years of the first, room-size electronic machines, smaller, faster devices based on transistors became available. Within 20 years, computers had powerful microprocessors. Today, computers are part of almost every home, school and business in the developed world and are built into machines that keep the modern world functioning.

LOGIC SWITCHES

All electronic computers are based on electronic switches or gates. Each one processes information using three basic logic operations—AND, OR and NOT. When these switches are arranged into huge, interlinked circuits, they can process many kinds of data very quickly. Early computers used lightbulb-size vacuum tubes (or valves) to build these logic circuits. Later models used smaller and more powerful electronic transistors, and eventually thousands of microscopic transistors were combined on the surface of one silicon chip.

Vacuum tube   Transistor circuit board   Silicon chip

Compact computing

Laptop computers house all the essential elements of the first electronic computers, but are over 200 times smaller and thousands of times more powerful. Through clever design, a laptop packs computer circuitry, keyboard, display screen, memory drives and power supply into a single, lightweight, portable package.

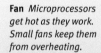

Central processing unit (CPU) *At the heart of the computer is the CPU—a data-crunching printed circuit on a tiny chip of silicon.*

Hard drive *Huge amounts of data are stored magnetically on the spinning, mirror-like disk of the hard drive. This allows the computer to store data even when switched off.*

Fan *Microprocessors get hot as they work. Small fans keep them from overheating.*

Battery *Laptops are powered by lightweight batteries that can last up to eight hours between recharges.*

TIMELINE
THE COMPUTER

c. 2500 BC Abacus
The ancient Babylonian abacus was little more than a set of pebbles shifted between marks on the ground or on a wooden board. Later, the Chinese developed more advanced counting devices using wires and beads.

1642 Mechanical calulator
French mathematical genius Blaise Pascal built the world's first adding machine to help his father calculate taxes. Using mechanical gearing, it could add numbers up to eight digits long, but could not subtract, divide or multiply.

1849 Difference Engine No. 2
Englishman Charles Babbage designed several "engines" that could perform complex calculations to 30 decimal place. These had many elements of the modern computer, such as memory, processors and programs.

Keyboard *The keyboard has its own microprocessor chip, which handles signals from switches beneath each key.*

Touch pad *The touch pad senses finger movement and pressure and converts it into an electrical signal. This avoids the problem of finding a flat surface for a mouse when on the move.*

OLED screen *Pixels within thin films of coloured organic molecules light up as they are supplied with electricity, producing a bright, colourful screen display.*

CD/DVD Drive *Text, audio, image and video files stored on CDs or DVDs can be accessed by a reader head and transferred back and forth between the hard drive and DVD drive.*

Motherboard *The motherboard houses the microprocessor and other computer circuits, connecting them to one another, to the power sources and to input and output devices.*

Mouse *A light-emitting diode (LED) and sensor tracks the movement of the mouse across a surface.*

1951 UNIVAC electronic computer
UNIVAC was the first commercially sold multipurpose computer. Its thousands of vacuum tubes, circuits and magnetic memory drums took up an entire room, and it often broke down.

1971 Microprocessor
Microprocessors cram thousands of circuits onto tiny silicon chips. They led to computers that were vastly smaller, cheaper and more powerful than previously thought possible.

Out of This World

Rocket

The rocket was first invented in China as a weapon, these early Chinese "fire arrows" were later developed into artillery by the armies of Europe. The idea of using a rocket to launch into space came about in 1926. American inventor Robert Goddard dreamed of sending manned rockets to the moon, but he realised that rockets with solid fuel were not powerful enough or easy enough to control for space travel. So he invented the first liquid-fuel rocket engine, and although he was ridiculed in the United States, the Germans took notice. During World War II, the Nazis used Goddard's designs to develop the V-2, a long-range rocket missile. After the war ended, the Russians and Americans used captured V-2 rockets to build the Vostok and Redstone rockets that would take both countries into outer space.

Vostok 1 "Swallow"

Though it was named after a small bird, the rocket that boosted Russian cosmonaut Yuri Gagarin into space was a giant. It stood more than 38 metres (125 ft) tall, and weighed more than 150 tonnes (165 t).

Orbiter *The Vostok command stage (or orbiter) came in two parts. The re-entry capsule contained the cosmonaut and the communications equipment. The equipment module housed retro rockets for turning the craft and gases for running the life support system.*

Communications antenna

Command control antenna

RE-ENTRY CAPSULE

Oxygen and nitrogen bottles for life support system

Communications antenna

Retro rocket

EQUIPMENT MODULE

Access hatch

Communications antenna

RE-ENTRY CAPSULE

Control instruments

Camera

CORE ROCKET

ORBITER

BOCTOK

Helmet

Spacesuit

Ejector seat

Porthole with optical orientation device

MISSION PROFILE

Like all rockets, Vostok suffered from a weight problem. More than 95 per cent of its weight was taken up by fuel, all of which was needed to accelerate the craft into space. The solution was to build it in three sections or stages, each of which dropped away after use. Only the re-entry capsule and Yuri Gagarin went to space and back.

1 06.07
Liftoff

2 06.09
Booster rockets jettisoned

3 06.12
Core rocket jettisoned

4 06.17
Final rocket jettisoned

5 06.17–07.35
In orbit

6 07.55
Gagarin ejects

7 08.05
Gagarin lands in a farmer's field

Re-entry capsule and ejector seat *The spherical re-entry capsule was barely bigger than Gagarin himself. The capsule could not land safely or gently; instead, the cosmonaut had to bail out in an ejector seat and parachute down to Earth.*

BOOSTER ROCKETS

TIMELINE

THE ROCKET

c. 1045 Chinese war rockets
The armies of China used exploding gunpowder to launch burning arrows and grenades at their enemies.

1944 V-2 rocket
During WWII, the Nazis launched about 2,900 explosive V-2 rockets at London and other targets.

1961 Vostok 1
Vostok 1 launched Yuri Gagarin into history on 12 April, 1961, making him the first man in space.

1981 Space shuttle
Reusable space shuttles were designed to make space flight cheaper and more routine. In total, five shuttles were built.

2004 SpaceShipOne
SpaceShipOne was the first spacecraft built by a private company. SpaceShipTwo, a passenger spacecraft, will soon follow.

Kindest Cuts
Surgery

Surgery is the branch of medicine that deals with treating diseases and injuries by operating on the body. Throughout history people have invented new methods and tools to cut, stitch, extract, replace and otherwise alter body tissue. Centuries ago, surgery often did more harm than good, but in the last 150 years improved knowledge and developments in tools and techniques mean we no longer have to fear the surgeon's knife. Anaesthetics and antiseptics help control pain, shock and infections. Syringes and catheters help drain fluids and deliver medicines. And thanks to endoscopic surgery, surgeons can operate inside the body through the tiniest keyhole incisions.

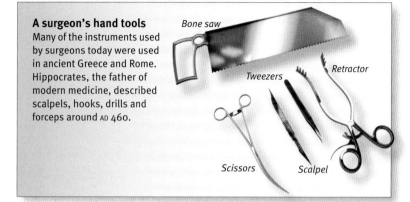

A surgeon's hand tools
Many of the instruments used by surgeons today were used in ancient Greece and Rome. Hippocrates, the father of modern medicine, described scalpels, hooks, drills and forceps around AD 460.

Bone saw

Tweezers

Retractor

Scissors

Scalpel

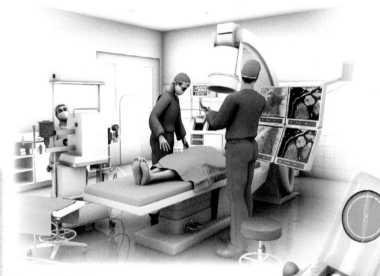

Operating theatre
In order to see what they are doing, endoscopic surgeons can track operations with video from a fibre-optic camera and live X-ray images called angiograms.

Pump it up *The surgeon inflates and deflates the catheter balloon using a handheld air pump. The pressure is precisely controlled with a syringe and can be released through controllable valves.*

GREAT MEDICAL INVENTIONS

In the wealthier parts of the world, people can expect to live long, healthy and mostly pain-free lives. This is due not just to advances in surgery, but to a long history of innovation in all branches of medicine.

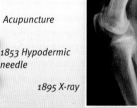

c. 2500 BC Acupuncture

1853 Hypodermic needle

1895 X-ray

1941 Antibiotics

1818 Blood transfusion

TIMELINE
SURGERY

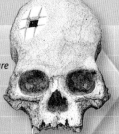

c. 7000 BC Trepanning
Trepanning—drilling a hole in the skull—is the oldest surgical procedure that we know of. It was generally thought to be a cure for severe headaches, epilepsy and insanity.

c. 2700 BC Ancient Egyptian surgery
Doctors in ancient Egypt trained for years in temple schools. They used knives, drills, hooks, needles, saws and forceps, and used stitches to sew up wounds.

1846 Anaesthetics
American dental surgeon William Morton demonstrated his anaest[...] ether inhaler in 1846. The news o[...] success—and his invention—spr[...] rapidly around the world.

Through the keyhole

Endoscopic—or keyhole—surgery makes use of the body's own pathways to reach a problem site and operate. This is preferable to opening up the body because it means less scarring, less chance of infection and faster healing and recovery. Illustrated on this page is an operation to treat a blockage in an artery that supplies blood to the heart muscles.

Catheter

Arterial wall

Stent

Way to the heart

To get to the operation site, the surgeon makes a tiny cut into the upper arm or thigh, and threads a thin tube called a catheter through the artery all the way to the heart.

Open wide

Once the catheter is in place, the surgeon can slide more tools through it and operate them remotely to relieve the blockage.

❶ Positioning *First, a thin tube carrying a deflated balloon is threaded through the catheter to the blockage site.*

❷ Inflation *Next, the balloon is inflated— which squashes the blockage, widens the artery and presses a thin, metal mesh tube called a stent against the artery walls.*

❸ All clear *Finally, the balloon and catheter are removed, and the stent remains in place to help keep the artery clear.*

1910 Endoscopic surgery
The first "keyhole" operation on a human being was performed in 1910. But this method has been perfected and used often only in recent decades.

1954 Organ transplants
The first successful whole organ transplant was a kidney transplant between twin brothers, performed by Boston surgeon Joseph Murray in 1954. Tissues like skin and bone had been transplanted earlier.

Access Control
Lock and Key

For almost as long as there have been treasures worth guarding, there have been locks and keys to secure them. In ancient Egypt, the poor kept their doors secured with simple wooden bars and latches, while the rich used complex wooden bolt locks with sliding keys. The ancient Greeks and Romans fixed their locks inside doors, and turned them with metal keys inserted through a hole. But these early inventions were easily outdone by clever thieves. It was only in the past 200 years that locksmiths invented the secure lock-and-key designs we all use. Today, engineers work with biometrics—physical features that make every person unique—and computer encryption to protect our property and information.

Eyeball entry
The iris is the coloured muscle surrounding the pupil of the eye. The patterns of the iris are unique to each person—even identical twins have different patterns. This makes the iris the perfect biometric key for use with computer-controlled security locks.

① Locked *Sprung metal pins in five slots jam the barrel in position and prevent it from turning.*

② Keyed in *The serrated edge of the correct key pushes the driver pins out of the barrel and aligns the tumbler pins.*

Driver pin

Tumbler pin

Spring-pin security
Each of the pin tumbler lock's five cylinder slots houses a sprung "driver" pin that pushes on a second "tumbler" pin. The pins are different lengths, which prevents the lock barrel from turning until the correct key is used to align them. The barrel is connected to a bolt that locks and unlocks the door.

③ Unlocked *The barrel is free to turn with the key.*

① Face up *A person looks into a digital camera that scans the face. A computer locates the eyes and a detailed image of the iris of one eye is captured.*

LOCK & KEY

c. 2000 BC Egyptian wooden lock
The ancient Egyptians locked their doors with bolts that were secured with wooden pins. To unlock the door, the pins were pushed up with a long wooden key.

c. 100 BC Warded lock and key
Warded locks were invented by the Romans. They have projections around the keyhole that prevent the flat face of the key from turning unless it has matching slots.

1778 Lever tumbler lock
The lever tumbler lock was the big advance in lock making for at lea 1,000 years. Only the correct key could move several levers, called tumblers, and open the lock.

2 Eye scan *The computer scans the image of the iris from the outside inward, identifying hundreds of reference points throughout the iris.*

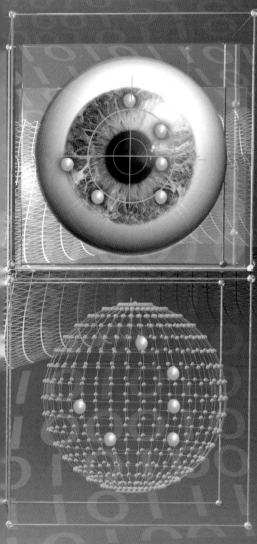

4 Searching *The computer searches through an existing database of irises, comparing codes and patterns until a match is found. Some systems can search millions of records in a second.*

3 Code making *The unique pattern of the iris is translated into digital code.*

5 Access approved *If the iris pattern is recognised and approved, the computer unlocks a secure room, building, bank vault or an encrypted computer file.*

1848 Yale lock
American Linus Yale designed an improved pin tumbler lock, which opened only after all five sliding pins inside were lifted into precise positions. His design is still in widespread use today.

1980s Biometric locks
The Chinese used fingerprints to identify people in the 1300s. Automated biometric security systems appeared in the 1980s. The first of these checked the shape of the hand or the pattern of blood vessels inside the eyeball.

Better Bodies
Bionics

Bionic devices replace or improve normal human body functions. Since the beginning of human history, those people left disabled by wars, accidents or birth have used artificial body parts to overcome their disabilities. Until the 1500s, anyone losing a limb had their wounds painfully sealed and were given simple wooden "peg" legs or metal hooks to replace them. French surgeon Ambroise Paré helped to change this by designing mechanical limbs to help disabled soldiers get back to a full life, rather than just get by. Since then, prosthetic—or artificial—limb technology has advanced rapidly. Bioengineers have not only built fully functioning bionic limbs, but have succeeded in restoring hearing and heart function with electronic implants.

Bionic leg
When we walk or run, we constantly adjust the way we move to keep balanced. The latest bionic legs use electronic sensors and computers to detect changing loads on the foot, ankle and knee and make the adjustments automatically.

Stepping out *Hydraulic components extend the leg as it swings through the air ready for the next step.*

HELPING HAND

Ambroise Paré (1510–90) was surgeon to four kings of France and a founding father of modern surgery and anatomical science. He was also a gifted inventor who designed prosthetic devices far in advance of anything else available at the time. His masterpiece was an artificial hand with movable fingers operated by gears, springs and catches.

Bionic foot *The best prosthetic feet do not rely on computers or motors—they are carefully crafted springs that return energy at the end of each step.*

TIMELINE

BIONICS

c. 300 BC Prosthetic limb
The oldest known artificial limb is a Roman wood-and-bronze leg dating from about 300 BC. Until recently artificial legs simply bridged the gap between stump and ground.

1923 Electric hearing aid
Austrian doctor Ferdinand Alt invented a crude electric hearing aid in 1906. His design was improved by adding vacuum tube amplifiers. Portable hearing aids were available from 1923.

1960 Cardiac pacemaker
American engineer Wilson Greatba[...] built the first cardiac pacemaker. H[...] invention delivers regular pulses of electrical current to the heart, keeping heart condition sufferers alive and well.

Bionic arm

Standard prosthetic arms are not connected to nerves, so are difficult to control and awkward to use. Bionic arms, however, can be controlled using the user's own nervous system. With this design, nerves leading to the missing arm are redirected to the chest, where electrodes attached to the bionic arm can pick up impulses and use them to direct arm movements.

Thought that counts *Although this man has lost an arm, the part of the brain responsible for movement still functions as if it were there. When he thinks about moving his left arm, nerve impulses are sent out just as they are when he wants to move his healthy right arm.*

Nerves *The nerves that would normally run towards the left arm have been surgically redirected to the muscles in the chest.*

Electrodes *Electrodes detect the nerve impulses where they connect with the chest muscle and relay them as electrical signals to the bionic arm.*

Movement *A computer processes the signals and directs the arm to perform certain movements such as bending the elbow, opening and closing the hand and extending the wrist.*

Radial nerve

Median nerve

Ulnar nerve

1969 Artificial heart
The first totally artificial heart was implanted in a patient in 1969. Even with current technology, they are a poor substitute for a transplanted donor heart.

1993 Bionic arm
Biomedical engineers in Scotland built the first robotic prosthetic arm in 1993. Now bionic arms can be precisely controlled, and even a sense of touch restored to the patient's existing nerves.

Wired World

Internet

The Internet has revolutionised human communication, knowledge and everyday life in much the same way that movable type did in the 1400s, or the telegraph and telephone networks did in the 1800s. The Internet began as a simple tool for computerised communication but has developed into a vast network that allows us to send messages, trade goods, be educated and be entertained. It connects us to a world of information, as well as a billion other users across the globe. Yet before 1960, computer modems could not connect more than two machines at once, and few people could have imagined a global network like the World Wide Web. It is thanks to the pioneering inventors of packet-switching technology, file transfer protocol programs and Web browsers, that we have the Internet we know today.

1 **File to fly** *Data—in this case a photo, but it could be an e-mail, Web page or computer file—begins its journey from one computer to another across the Internet.*

2 **Packets** *The photo file is split into multiple pieces called packets. Each packet carries the receiver's Internet protocol (IP) address, which identifies the exact computer that the file must reach.*

3 **Routing** *Routers determine the best path for the packets to take through the network. A packet will usually be directed by many routers on its journey and will often take a different path from other packets in the same message.*

4 **Reassembly** *When all the packets have arrived at the destination computer, the file is reassembled. The whole process usually takes no longer than a few seconds.*

MAPPING THE INTERNET

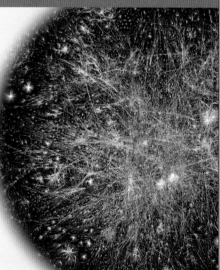

This map of the Internet represents the dizzying maze of connections among hosts, servers and routers. Seen as a whole, the Internet looks almost like a nervous system or brain—and that may be no accident. Like the brain, the Internet has billions more connections than it actually uses and signals are not restricted to a single route through the network.

Packet-transfer puzzle

What ultimately makes the Internet work are packets and routers. An e-mail message, photo or any other type of file does not make its way through the Internet in one piece. Instead, it is split up into small pieces of data called packets. Routers are computers that identify each packet's end destination and send them through the best available pathway. Packet splitting helps to avoid glitches and keeps the data flowing freely, as single packets are easily rerouted around broken network connections.

TIMELINE

THE INTERNET

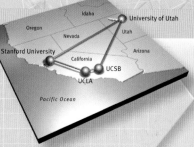

Idaho
Oregon
Nevada
University of Utah
Utah
Stanford University
Arizona
California
UCSB
UCLA
Pacific Ocean

1969 ARPANET
The Internet began life in the ARPANET computer network, designed by the United States military's Advanced Research Projects Agency (ARPA). The first four nodes were linked in December 1969, and by 1975, ARPANET had more than sixty.

1971 E-mail
Programmer Ray Tomlinson created computer codes that assigned addresses to different users and machines. He sent the first e-mail between two computers using the ARPANET. Thirty years later, over a billion e-mails a month were being exchanged over the Internet.

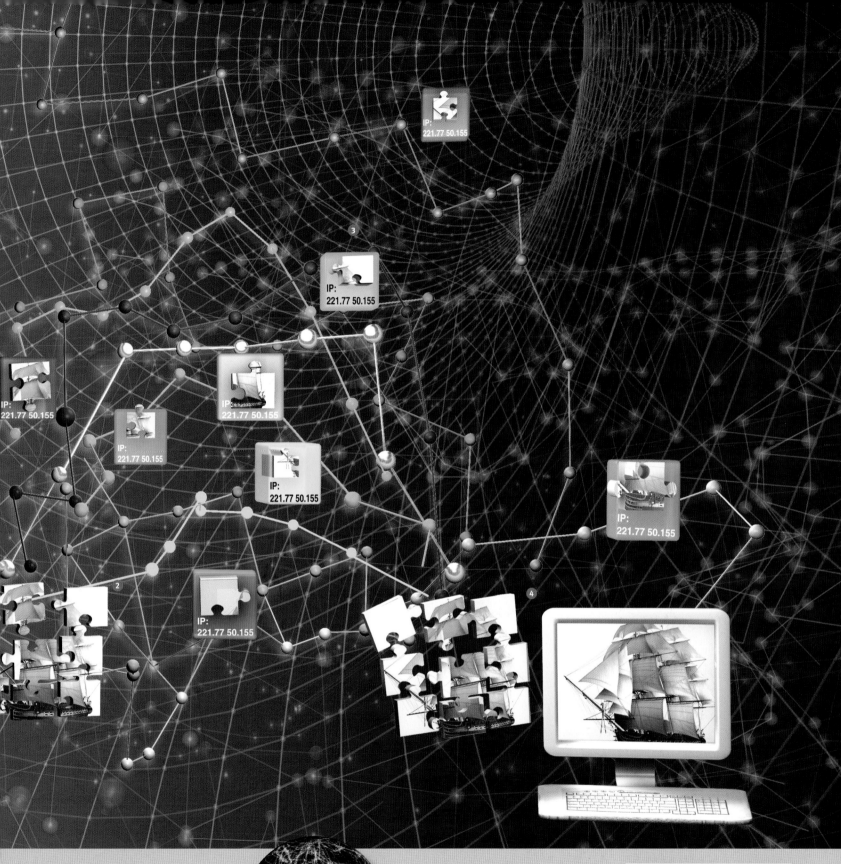

1977 Internet

In 1977, programmers Vint Cerf and Robert Kahn demonstrated their newly invented transmission-control protocol (TCP) by "inter-netting" (or linking) three distant computer networks in San Francisco, Virginia and London into one. The global Internet was born.

1991 World Wide Web

Tim Berners-Lee, a researcher at the CERN particle physics laboratory in Switzerland, invented hypertext and the first Internet browser to help manage information and data from experiments. His "World Wide Web" system shaped the Internet we use today.

Glossary

AD An abbreviation for the Latin *anno Domini*, meaning "in the year of our Lord". Used for the measurement of time, AD indicates the number of years since the supposed date of Christ's birth.

aerodynamic The streamlined shape of a vehicle, designed to cut down on drag from the air moving around it. This is especially important in aircraft design.

aeronaut A pilot or other crew member of a lighter-than-air craft, such as a balloon or airship.

airliner A large aeroplane used to carry passengers rather than cargo.

air pressure The force of the air inside a container or in Earth's atmosphere.

alloy A material made of two or more different metals, or a mixture of metals and nonmetals.

amplification The process of making a sound louder or a signal stronger, often using mechanical or electronic devices.

anaesthetic A drug that stops the body from feeling pain and other sensations.

analogue A machine or device that uses variable inputs and outputs (as opposed to a digital device, which turns digital or variable inputs into numerical or number-coded outputs).

antenna A device that receives or transmits radio waves. An antenna connected to a radio receiver changes radio waves into electrical signals. An antenna connected to a radio transmitter changes electrical signals into radio waves.

antiseptic Chemical substance that kills bacteria and other harmful micro-organisms, preventing infection—especially of open cuts and wounds.

artillery Military weapons used to fire heavy shells or exploding missiles.

assembly line Part of the mass-production method of manufacturing. Workers fit one part to a product as it moves past them on a conveyor system.

astrolabe A navigational instrument used by early sailors to measure the position of the stars and planets to find out how far north or south they had travelled.

automated Done partly or completely automatically, with the assistance of machines—reducing the need for human work.

axis The point around which something rotates.

axle A rod, bar or pole that holds a wheel or pair of wheels.

BC An abbreviation for "before Christ". Used for the measurement of time, BC indicates the number of years before the supposed date of Christ's birth.

biotechnology The process of changing or controlling living things to make new products.

biplane An aeroplane with two sets of wings—one positioned above the other.

calculator A machine that can add, subtract, multiply and divide numbers.

celluloid A tough material originally used for motion picture photography.

clockwork The cogs, wheels, gears, springs and shafts used to make mechanical clocks work.

code A system of symbols where each symbol represents a different piece of information.

computer A machine that automatically performs calculations according to a set of instructions that are stored in its memory.

cosmonaut A person trained to pilot or travel in a spacecraft. An alternative word to "astronaut", used by Russian and Soviet space programs.

diaphragm A thin disc of flexible material, used to pick up or create sound waves.

digital Information expressed as a series of numbers, such as computer binary code, which uses a series of 1s and 0s to represent numbers, letters, images, sounds and video.

ejector seat An emergency escape seat that launches from a crashing aircraft or spacecraft and parachutes the pilot to safety.

electricity The movement or flow of charged particles used as a source of energy for electrical devices and machines.

electrode An electrical conductor through which an electric current enters or leaves an electrical circuit into or from another medium.

electromagnet A magnet made by wrapping a coil of electrical wire around an iron object, which can be switched on and off with the electric current.

electromagnetic wave A wave of energy made of vibrating electric and magnetic fields. Light, radio and X-rays are examples of electromagnetic waves.

electrostatic charge Energy produced or caused by static (or nonflowing) electrically charged particles.

engineer A person who designs or builds engines or machines.

entrepreneur A person who starts new—and often risky—businesses.

generator A machine that converts energy from one form to another—usually from movement to electricity.

girder Supporting strut or beam used to build strong frameworks—usually of bridges, buildings or large aircraft.

gravity The force of attraction towards Earth that keeps everything on the ground.

gunpowder A mixture of potassium nitrate, sulphur and charcoal used as an explosive and in fireworks.

horsepower Unit of power often used to describe engine power. Roughly equal to the pulling power of an average horse.

hydraulic Fluid pressure used to move, control or operate a machine.

hydrogen The lightest chemical substance in the universe, which normally takes the form of a flammable gas.

industrialisation The process of introducing machinery to help produce or make goods. Countries that use machines this way are said to be industrialised countries or nations.

infrared Type of radiation that sits between light and radio waves on the electromagnetic spectrum. Infrared radiation is invisible to the naked eye, but is often used in night-vision goggles and remote-control devices.

laser An intense light of one wavelength and frequency that can travel long distances. It is used to cut materials, carry television transmissions, print onto paper and guide machines.

locomotive A self-powered vehicle that runs on a railway track.

mass production A method of manufacturing large quantities of goods, often using a number of machines. Each worker or machine in a factory works on just one part of a product.

mechanism The parts of a machine that come together to make it work, or the arrangement of parts within a working machine.

medieval period The period in Europe between about AD 500 and 1500.

microwave An electromagnetic wave with a frequency longer than infrared but shorter than radio waves. Microwaves are used for radar, to transmit data and to cook food by exciting water molecules.

modem Electronic device used to send digital information between computers through telecommunication cables.

optical fibre A communications cable made from solid glass or plastic fibre. It transmits light from one end to the other, even when the cable is curved or bent.

pendulum A swinging weight used to regulate the timing and motion of mechanical clocks.

physicist A scientist who studies the nature of forces, motion, matter and energy.

piston A movable, solid cylinder that is forced to go up and down inside a tube by the exertion of pressure.

pneumatic A machine that operates using compressed air, or an object filled with pressurised air.

propeller A set of rotating blades used to propel a boat or aircraft forward by pushing water or air backwards.

prototype The first version (or an early working model) of a machine or invention.

radar Electronic device used to locate an object or vehicle by sending out radio waves and detecting those that bounce off it.

radio waves Invisible electromagnetic waves that carry information such as Morse code "beeps" and the human voice.

Renaissance The period in Europe between 1300 and 1500 when science, invention, art and education were strongly encouraged.

retro-rockets Small rocket engines, separate from main rocket engines, that manoeuvre or slow down a spacecraft in space using short blasts of gas in the opposite direction.

shutter The part of a camera that opens for a fraction of a second when a picture is taken to let light fall on the film or sensor behind it.

silicon A common substance found in sand and clay. It is used in computer chips and solar cells.

stator The part of a rotary machine that does not move.

stereo Type of sound reproduction in which two microphones and/or speakers are used to create a three-dimensional effect. Also a short name for a stereophonic record player.

technology The tools and methods for applying scientific knowledge to everyday life.

telegraph The earliest system for communicating at a distance by electrical wire. Telegraph messages were communicated by simple on-off–type electrical signals.

thrust The force of a jet engine or rocket engine that drives the engine forward.

turbine A wheel with any blades that is made to turn by a gas such as steam, or a liquid such as water. It is used to power machines or to generate electricity.

type A piece of rectangular metal with a raised letter or symbol on one side.

valve A device used to regulate the flow of a gas or a liquid, or to turn it on and off.

X-rays Electromagnetic waves that can pass through soft parts of the body. X-rays are used to create images, on photographic film or computer screen, of the inside of the body.

Index

Credits

The publisher thanks Alexandra Cooper for her contribution, and Puddingburn for the index.

ILLUSTRATIONS
Front & back cover Malcolm Godwin/Moonrunner Design; **spine** Godd.com **Leonello Calvetti** 18–19; **Malcolm Godwin/Moonrunner Design** 8–9, 12–13, 20–1, 24–5, 28–9, 32–3, 40–1, 42–3, 46–7, 56–7, 58–9, 60–1; **GODD.com (Markus Junker, Rolf Schröter, Patrick Tilp)** 16–17, 22–3, 26–7, 30–1, 34–5, 36–7, 38–9, 44–5, 48–9, 50–1, 52–3, 54–5

PHOTOGRAPHS
Key t=top; l=left; r=right; tl=top left; tcl=top centre left; tc=top centre; tcr=top centre right; tr=top right; cl=centre left; c=centre; cr=centre right; b=bottom; bl=bottom left; bc=bottom centre; bcr=bottom centre right; br=bottom right

CBT=Corbis; GI=Getty Images; iS=istockphoto.com; MEPL=Mary Evans Picture Library; NASA=National Aeronautics and Space Administration; PL=photolibrary.com; SPL=Science Photo Library; SS=Science and Society Picture Library

8cl iS; **10**tl iS; l PL; c CBT; **16**b PL; **17**bl, br iS; **18**tcr, tr iS; **19**t, br iS; tr PL; **21**b SS; bl, br iS; **24**c CBT; cl PL; **25**bl SPL; br iS; **26**cl PL; **28**c iS; **26**br NASA; **29**bl GI; **30**cl CBT; **31**br iS; **32**b CBT; bl PL; **33**br GI; **34**bl iS; **35**bc, br iS; **37**tr CBT; **39**bc iS; **41**bl MEPL; **42**bl PL; cl CBT; **43**bl PL; **45**br iS; **46**br PL; **49**bl GI; br iS; **50**t PL; cl CBT; bl iS; **53**bcl PL; **54**b PL bcr; bl, br, cl, cr iS; **55**br PL; **57**bc, br iS; **58**bl MEPL; **59**bl PL; bc CBT; **60**l The Opte Project; br iS; **61**bl, br iS

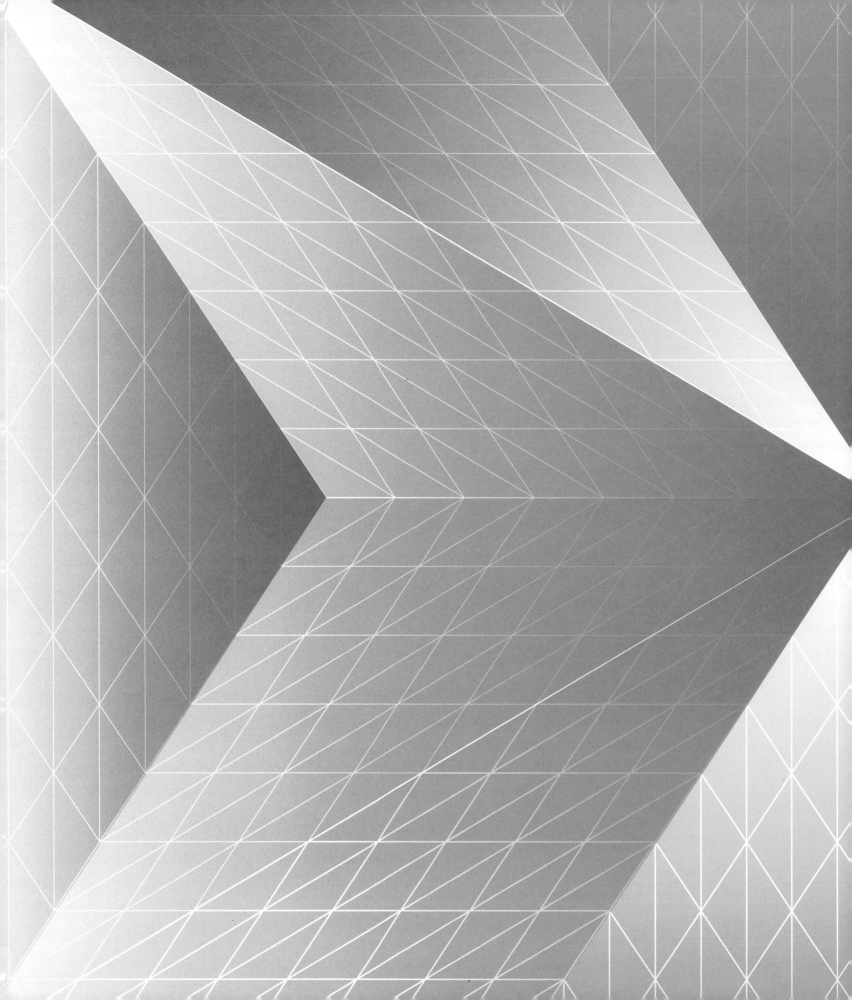